Ramiro Luz

Coding Dojo for teaching agile practices

Ramiro Luz

Coding Dojo for teaching agile practices

A participatory and inclusive method for teaching agile development

ScienciaScripts

Imprint

Cover image: www.ingimage.com

This book is a translation from the original published under ISBN 978-3-330-75775-2.

Publisher:
Sciencia Scripts
is a trademark of
Dodo Books Indian Ocean Ltd. and OmniScriptum S.R.L publishing group

120 High Road, East Finchley, London, N2 9ED, United Kingdom
Str. Armeneasca 28/1, office 1, Chisinau MD-2012, Republic of Moldova, Europe
Printed at: see last page
ISBN: 978-620-8-29178-5

SUMMARY

SUMMARY

LUZ, Ramiro. The influence of the Programming Dojo on the teaching of agile practices. 57 f. Dissertation - Postgraduate Program in Applied Computing, Federal Technological University of Paranà. Curitiba, 2017.

Programming Dojo is a dynamic and collaborative activity inspired by martial arts where you can practice programming, especially techniques related to agile methods. Current teaching methods treat all students in the same way. Each person has a different background, experiences, skills and knowledge. The particularities of each individual are rarely respected. We sought to verify the influence of the Programming Dojo as a teaching activity in order to promote interaction between students. Does the Programming Dojo offer the necessary re-courses for the academic teaching of computer programming? To answer this question, questionnaires were carried out with laypeople and active participants in Programming Dojo, spontaneous participants in Programming Dojo groups and compulsory participants of students who took part in Programming Dojo in a regular specialization class at the Federal Technological University of Paranà. Interviews were conducted with experts who organize Programming Dojo meetings. Positive points, such as student participation, and negative points, such as the fact that it is not a suitable activity for presenting theoretical subjects, were weighed up. The author concludes, based on the answers to the questionnaires and the opinions expressed by experts in the interviews, that the activity can be used in a complementary way in computer programming courses in order to increase student participation and allow the teacher to get to know the difficulties and facilities of each individual student.

Keywords: Programming Dojo, TDD, Extreme Programming XP, programming education, agile methods, pair programming, interactive learning environment, collaborative learning, problem-based learning

1 INTRODUCTION

Dojo, translated from Japanese, means "place of the path". This term was originally used to designate the meditation space of Buddhist monks. In this context, it can be translated as "place where life is studied". Today it is known as the place where martial arts are practiced.

Martial arts practitioners are familiar with the term *Kata. Kata* is a sequence of exercises repeated several times with a small improvement with each exercise. In his *blog*[I] , Dave Thomas presents the idea of code *Katas* (THOMAS, 2007a), realizing the need for programmers to practice just as athletes and artists do. Musicians, for example, repeat the same piece of music over and over again in search of improvement; they are exercises designed to be carried out with the aim of learning and reinforcing concepts during practice, without the commitment of achieving a single correct answer.

Unlike the problems proposed in programming marathons where the focus is on competition, such as (ACM International Collegiate Programming Contest.,), Code *Katas* are used as an instrument for collaborative exercise. Eventually, challenges similar to those found in (SKIENA; REVILLA, 2003) can be used, however it is important to reinforce the idea that Code *Katas* or challenges, used during a Programming Dojo are simple problems that serve as a tool for collaborative and inclusive learning, competition is discouraged during the Programming Dojo.

There are a few variations in the way of conducting a Programming Dojo. The *Prepared Kata* style, the *Randori Kata* style and the *Kake Kata* style.

The style called *Prepared Kata* consists of a simulated presentation of a Programming Dojo, usually held at events to publicize the Programming Dojo to relatively large audiences. It can be interrupted at any time to clarify any questions the audience may have. It also allows those interested to reproduce the steps afterwards.

The most widespread and used style, *Randori Kata*, is initiated by choosing a problem, challenge or *Kata*. The *Kata* will be used as a means of exercising some agile practices such as pair programming, test-driven development and baby steps on a single computer connected to a projector. It has a mechanism for changing the pair, there are several mechanisms, but the most commonly used is called *TimedBox*. At each short interval of time, between 5 and 7 minutes for example, the pilot returns to the audience, the co-pilot takes over the keyboard and a person from the audience assumes the role of co-pilot. You should always start from scratch, even when a problem has already been used in a meeting, everything that has been done should be ignored and a new solution started. The practices of test-driven development and baby steps should be used. All participants are expected to pay attention to what is being done. The pair should constantly explain what they are doing so that everyone can follow along. The audience can make suggestions when the tests are passing, this state is called the green light. The pair must not be interrupted when the tests are not passing, this state is called a red light. The duo can ask the audience for help. The Dojo ends when the time runs out, usually after two or three hours of work. About 20 minutes should be set aside for a retrospective of the activity, highlighting what could

Ih ttp://codekata.pragprog.corn/

be improved and what was good.

The *Kake Kata* style follows the same principles as the *Randori Kata* style, with a few differences. Several computers are used, in at least one variation. The pairs work on different problems, with different languages, different environments, and it can even happen that everything is different. Those who don't take part in any of the pairs shouldn't get in the way and are free to wait for the pairs to change. After the time shift, the pilots must return to the audience, or they can become co-pilots on another computer. The co-drivers become pilots and members of the audience can become co-drivers. At the end, in the same way as the *Randori* style, a retrospective should be held.

Dave Thomas realized the advantages of practicing without time pressure, through small incremental steps, one day when he had some free time while waiting for his son at karate class. By challenging a problem without commitment, he realized that this is a good way to practice and improve. In another conversation with a friend about the practice of musicians, he saw the possibility of adapting the idea for programmers, who rarely practice outside the workplace. In this way, the practice of programming with *Code Kata* was conceived (THOMAS, 2007b). The principle behind this idea is deliberate practice (ERICSSON et al., 1993).

According to (ERICSSON et al., 1993), the maximum level of performance for an individual in a given domain is not obtained automatically as a result of long experience, but the level of performance can be improved in highly experienced individuals as a result of deliberate efforts to improve. He also states that in the absence of adequate *feedback*, efficient learning is impossible and improvement is only minimal even for highly motivated subjects. (ERICSSON et al., 1993), explains that, considering three types of generic activities, working, playing and deliberate practice. Working includes public performances, competitions, services rendered for payment and other activities directly motivated by external rewards. Play includes activities that have no explicit goal and are inherently enjoyable. Deliberate practice includes activities that have been specially designed to improve the current level of performance. The goals, costs and rewards of these three activities differ, as does the frequency with which individuals pursue them. It suggests, then, that deliberate practice as an effort activity can be sustained for a limited time each day over extended periods without leading to exhaustion. To maximize the long-term gains from practice, individuals need to avoid exhaustion and should limit practice to an amount that they can fully recover from on a daily or weekly basis.

These ideas gave rise to *Coding Do*jos, which are, in a nutshell, gatherings of programmers with the intention of practicing programming using agile methods. The Coding Dojo provides an appropriate environment for deliberate effort. It has mechanisms that allow the participant to make a self-assessment and for other participants to give the necessary *feedback* with the aim of continuous improvement. The Programming Dojo has activities specially designed to improve programmers' skills and is held on a weekly, fortnightly or monthly basis, thus avoiding exhaustion.

The activity, Programming Dojo, is disciplined and can best be explained by drawing an analogy with martial arts. During the practice of martial arts, an instructor, usually called a *sensei*, organizes the disciples in a wide circle and then the *sensei* invites two disciples to the center. They fight according to the rules of the sport *and* the *sensei*'s instructions. After a few minutes of fighting, the disciples leave the center of the circle, sit down and the *sensei* invites another two disciples to the center. In this way, the disciples who are fighting learn by practicing, the disciples who are in the circle learn by observing and the more they fight and observe, the better they learn.

In the Programming Dojo, participants collaborate with each other and everyone learns together. Figure 1 shows a photo of a Programming Dojo meeting. In simplified terms, it can be said that during a Programming Dojo meeting, programmers come together to solve programming challenges. The challenges must be solved using agile method practices. On a single computer, participants take turns in pairs to solve a challenge, always taking care to explain to all participants what is being done. The computer screen is projected so that everyone can see what is happening.

Figure 1: Photo from a Programming Dojo meeting

Source: Author

To hold Programming Dojos, you need a projector, a computer and a board on which to write or draw something related to the activity. Preferably, the room should not have computers so that everyone can pay attention to the pair that is programming and the projected solution. Optionally, a snack is served after the activity, providing a moment of socialization and relaxation where discussions can continue.

Each meeting is divided into stages (GAILLOT, 2012). There may be variations from group to group, but the most common is to follow the process suggested by the first known Programming Dojo (BOSSAVIT; GAILLOT, 2005b).

The first thing to do is to set a date for the next meeting, so that participants can organize themselves to take part in the activity. In some groups, the meetings always take place on the same day and at the same time. It is suggested that you make this decision in about two minutes.

The next step is to look back over the last session, discussing what was frustrating, what was interesting and what went well. This stage lasts about twenty-five to thirty minutes. Some groups do the retrospective right after the end of each meeting. The aim of this stage is to carry out a self-evaluation in order to improve future meetings.

Participants then decide what the topic of the session will be, what problem and what programming language will be used to address the problem. Lasting around ten minutes, this stage prioritizes the choice of a technology known to at least one participant. It is not possible to carry out the activity if no one can answer questions and give advice on the best techniques and operation of the tools available for the chosen programming language.

For around forty minutes, participants program using agile development techniques and following the style of the Programming Dojo adopted, *Prepared Kata*, *Randori Kata* or *Kake Kata*.

Optionally, a break of five to ten minutes can be taken to reflect on the progress and approach of the solution; some groups do this throughout the development and others do not. After the break, the development continues for another forty minutes.

Some agile method techniques receive more attention, especially test-driven development, pair programming and baby steps.

Test-driven development (TDD) is an agile practice based on the production of automated tests that verify the functionality of a piece of software. The process of testing and creating functionality guides design and development. For each small piece of functionality, a test that describes and verifies what the code should do is written. Only the code that is necessary and sufficient for the code to pass the test is then produced. This is followed by refactoring, a term used for improvements that simplify and make the code clearer, of the functionality and test code. According to (BECK, 2002), this allows reliable programs to be written, with unit tests and simple designs for complex problems. Tests provide constant feedback on the functioning of the code and provide security for changes. When something needs to be changed, it is possible to make the change and run the automated tests, which will check everything that was already working. This means that the developer doesn't have to keep evaluating the code to see if a change will affect other parts of the software that were already correct. The development sequence used in TDD is as follows:

- First, you write a simple test for a small step in solving the problem.
- You should run the tests to see what failed, some tools already do this automatically.
- Then you write the minimum code necessary for this test to be correct.
- Tests are run to confirm that it works, and the term most commonly used by participants is: the code has passed the tests.
- So if the code passes the tests, you can improve it or, in other words, refactor it.
- And finally, a new test is written for a new stage, going back to the beginning.

During the test-driven development cycle, another agile practice, called baby steps, recommends keeping things simple, without going too far at a time (BECK; ANDRES, 2004). This practice provides a gradual advance in small steps, thus increasing the chances that the next step will be trivial (BOSSAVIT; GAILLOT,).

Pair programming, one of the most important agile practices in the Programming Dojo, consists of two developers sharing the same computer. The pair collaborates continuously on the same project, algorithm, code or test. This practice encourages constant communication and one of its main advantages is instant code review. One of the members of the pair, called the pilot, uses the keyboard to code. The other member of the pair, called the co-pilot, observes the pilot's work looking for errors in syntax, typing, calling the wrong function or procedure and so on, (WILLIAMS; KESSLER, 2002).

In Programming Dojos, participants play roles. One of the participants is the Sensei, who organizes the activity. The "audience" observes the evolution of the code and the "pair" develops the code by practicing pair programming, one of them as the "pilot" and the other as the "co-pilot". Practices of agile methods, TDD, refactoring, baby steps (BECK; ANDRES, 2004) and (BOSSAVIT; GAILLOT,) and pair programming should be applied. In this way, and over time, participants become familiar with TDD techniques and learn from each other in a collaborative and evolving way in a friendly environment (SATO et al., 2008).

This research consisted of questionnaires with participants in Programming Dojos with different backgrounds, some more experienced, some volunteers and some students who participated compulsorily during regular classes. Interviews were held with experts who are experienced in running programming dojos.

1.1 MOTIVATION AND JUSTIFICATION

As reported in (PIMENTEL et al., 2003), there is a low rate of student assimilation in subjects that require knowledge of programming. One of the reasons for this, according to (PIMENTEL et al., 2003), is the difficulty in dealing with differences in background, experience and skills between learners. Teachers and current methods do not always make it possible to identify each individual's problem. These difficulties can be diagnosed by the high level of repetition in introductory subjects and by the difficulties shown by students in advanced subjects that require the prerequisite of programming, he says (PIMENTEL et al., 2003).

We then sought to identify the influence of the Programming Dojo as a teaching activity in order to promote interaction between students and help the teacher to identify the difficulties of individual students.

1.2 OBJECTIVES

The Programming Dojo presents some characteristics similar to problem-based teaching (PRINCE, 2004) in a participatory and collaborative environment. The possibility of using this activity to teach agile practices guided the following objectives.

1.2.1 General Objective

Verify the influence of the Programming Dojo in programming teaching environments using agile practices.

1.2.2 Specific Objectives

In line with the general objective, the following specific objectives were determined.

- Get the opinion of experts and laypeople on the practice of the Programming Dojo.
- Identify the advantages and disadvantages of the Programming Dojo as a teaching activity.

- Compare the answers obtained from different groups and check the participants' positive or negative perception of the activity.

2 LITERATURE REVIEW

In search of similar work, the author carried out a bibliographic survey using the search tools in the main academic article databases, filtering out articles from the exact sciences area. Publications were selected that evaluated the teaching of agile methods or using agile methods in teaching as well as publications related to Programming Dojo.

1.1 RESEARCH INVOLVING AGILE METHODS, TEACHING AND EDUCATION

In the article (SCHNEIDER, 2005), the authors examined the educational goals for undergraduate software engineering education and how these goals could be met with Extreme Programming (XP) practices. The authors pose the question: Is it possible to make the teaching of software engineering more practice-oriented and address the misconception that the main concern of software engineering is to produce one document after another? A discussion of Extreme Programming in an educational environment is presented. According to the authors, Extreme Programming (XP), despite having many positive aspects, some of its principles are insufficient for the learning needs of most students. They conclude by stating that Extreme Programming (XP) has limited value for large-scale systems development education. Limited resources in many institutions make it questionable whether the necessary infrastructure can be set up in order to apply Extreme Programming (XP) properly. While some practices are suitable for the educational environment, others contradict the educational goals. In particular, the focus on code makes it difficult to introduce concepts that are necessary for the development of large-scale systems. On the other hand, some Extreme Programming (XP) practices can be useful for teaching low-scale systems development, such as Pair Programming.

During a software engineering course at Meredith College, an experiment was carried out by Barrett Koster (KOSTER, 2006). During the experiment, students realized the benefits of pair programming and were able to learn in a collaborative and collective environment instead of the traditional competitive and individualistic environment. Through the possibility or need to share code, they learned to create well-documented code, encapsulated appropriately, so that it could be integrated with other students' code. Concepts of good programming practice were reinforced: object orientation, file organization and versioning, well-written comments, standardization, nomenclature and modularization are some examples that were seen in practice. Another positive point, related to TDD, was that the students realized that the tests helped to clearly define the goals that had to be achieved once the test cases had been written.

The interactions allowed the students to evolve the code little by little by having working parts throughout the phases, which reduced the frustration and feeling of failure often seen in computer science courses. The authors concluded that agile methods have helped to improve the teaching of software engineering.

In previous classes, students complained about their teammates and many students had poorly functioning programs. They were only concerned with passing. The contrast was obvious: everyone had working programs and there was a feeling of camaraderie in the lab. According to the authors, it was the first time that the projects had been fully tested. In the end, they report that agile methods have helped to improve the teaching of software engineering.

Agile methods generally focus on people and their behavior. In the study carried out by (MELNIK; MAURER, 2005), undergraduate students who were invited voluntarily were evaluated. They answered questionnaires about their opinions and experiences with agile methods. Indications were found that the students are very enthusiastic about agile practices. Indications of the development of professional qualities were seen in team members working with agile methods, such as communication, commitment, cooperation and adaptation. The authors state that future work could be carried out to validate the results.

The article (MILLER; SMITH, 2007) evaluated the use of test-driven development in embedded programming classes. Teaching embedded programming is complex for a number of reasons. The development environments are different from what students are used to. James Miller and Michael Smith suggest in their article (MILLER; SMITH, 2007) that using TDD to teach beginner students embedded development can be an excellent tool to combat this complexity. The authors believe that instructors need to adapt teaching mechanisms to reduce the difficulties, proposing the adoption of TDD as a central component of this process, giving the following reasons:

- Specification by example;
- Automatic *feedback;*
- Specification and incremental development;
- Formal approach to testing;
- Non-functional specifications;
- Non-functional tests;
- Solid foundations for refactoring non-functional requirements.

The authors received positive feedback from some students. Students appreciated the ability to use existing tests to verify that the equipment is working outside of regular lab hours, while others showed a lack of interest in continuing with TDD. The students who evaluated positively wanted to know how to encourage other teachers to adopt TDD. The article deals with the authors' experience; at the time of publication, the work was still in progress.

In another work, a real case study, students learned agile practices through a project managed using the Scrum method (BRUEGGE et al., 2009). The authors report that the project was a success and an example of the application of agile practices and principles, showing how agile ideas can be taught in academic environments. They also mentioned that the students had fun while learning and developing the project.

In the article (VODDE; KOSKELA, 2007), the authors report on their experience creating test-driven development courses at Nokia Networks. According to the authors, the first course was essentially a series of lectures that didn't have the expected impact. So they revised the course and adapted an exercise for counting lines of code described in Dave Tomas' Katas of Code (THOMAS, 2001). The authors say that by following test-driven development and refactoring the code, they had to face the facts and take a stand on the poor architecture. They concluded that the line-counting exercise is extremely useful in providing insights into test-driven development and its benefits, including:

- Removing external dependencies helps improve testability.
- Reflective thinking promotes the improvement of solution architecture.
- A well-designed architecture and good test coverage also help new architectures to emerge.

They conclude by commenting that the lesson learned from conducting the course with this simple exercise is just one of those they have learned.

(SINIAALTO; ABRAHAMSSON, 2007) evaluated projects developed using TDD and found that TDD does not produce highly cohesive systems when used by amateur developers. It was also observed that TDD does not automatically produce cohesive code. The mere fact of using TDD is not synonymous with quality: training is required.

One of the most important activities in the Programming Dojo is Pair Programming. Research on Pair Programming used in undergraduate courses was found as follows.

In the article (BALIJEPALLY et al., 2009), the authors report on research carried out with students to evaluate the efficiency of pair programming. Some students worked in pairs and others individually, but their performance was measured in nominal pairs, pairs of students programming separately. The students solved introductory tasks of high and low levels of complexity. Software quality, satisfaction and confidence in performance were measured.

The authors pose the question "Is pair programming better than individual programming?". The aim was to study the effectiveness of pair programming compared to the effective performance of individual programmers. To this end, they carried out a controlled experiment in a laboratory: students worked on introductory level tasks, where two factors were manipulated: programming mode (pairs versus individual) and task difficulty (high and low complexity). Individual participants were combined into nominal pairs. The performances of the collaborative pairs were compared with the two best nominal pairs.

A combination of results forms the basis for the answer to the question: software quality, satisfaction and confidence in performance. The limitations reported were: the participants' reported experience was 2 years or less (38% reported 2 years of experience); the students might not remember the syntax of the programming language (access to online documentation was provided); the difference between the workplace and the laboratory environment, among other factors. According to the authors, the performance of pair programming does not exceed the performance of its best member working individually. No obvious effect was observed with regard to the complexity of the tasks. Pairs felt more satisfied and confident in their performance.

1.2 PUBLICATIONS RELATED TO PROGRAMMING DOJO

In (SATO et al., 2008), they present the concepts of the Programming Dojo, the lessons learned during the weekly meetings at the University of São Paulo - USP and discuss the aspects of the Programming Dojo that promote learning and knowledge sharing.

The article (ANICHE; SILVEIRA, 2011) assessed learning in environments where teams use agile methods. Through semi-structured interviews with employees, their opinions on learning in the company where the authors work were evaluated. One of the activities was the Programming Dojo. According to the employees, they did not perceive any evolution due to the baby steps (BECK; ANDRES, 2004), possibly caused by the fact that the team learns easily and quickly. Alternatives were tried, changing the format of the Programming Dojo, instead of using the *Randori Kata* format, they used the *Prepared Kata* and *Kake Kata* formats. What was a consensus among the collaborators was that learning by programming in pairs was mentioned as a way of enhancing learning.

Agile methods techniques, like other programming techniques, are taught in practice. It is believed that, due to the nature of the activity, this is the best way to teach and learn. In the article (BRAVO; GOLDMAN, 2010), the evaluation of two methods for teaching agile methods is presented. They evaluated the Programming Dojo, through questionnaires the participants suggest that the activity is very effective, the other method used self-modeled tools, which showed little influence on the teaching of agile methods. The article was based on his dissertation (BRAVO, 2011). Participants answered online questionnaires about their perception of learning voluntarily. Indications were found that practice is a good teaching technique for those with little knowledge or experience of TDD, but the effect on more experienced participants is unclear.

Laurent Bossavit and Emmanuel Gaillot presented a Workshop on Programming Dojo (BOSSAVIT; GAILLOT, 2005a). During the workshop, the mechanisms of the Programming Dojo were recreated so that participants could experience the activity and eventually be inspired to start their own Programming Dojos. Emily Bache presented a similar workshop with a slight variation in (BACHE, 2009).

In (CARMO; BRAGANHOLO, 2012) the authors studied different forms of Programming Dojo and evaluated the acceptance of students on undergraduate courses. Reports were collected from students claiming motivation problems, shyness and insecurity as reasons for not taking part in the Programming Dojo. The dispersion of the audience caused discomfort during the activity in one of the formats. In another format, lack of communication was reported as a problem. The students showed a preference for laboratory classes. They suggest that the difference between a traditional Programming Dojo, where participants attend voluntarily, and the adapted Dojo, where students are obliged to participate, is a factor that hinders student motivation. In addition, points were distributed to the team with the best performance, favoring student motivation. According to the authors, the students considered the first version of the Programming Dojo, called Individual Dojos, to be the worst approach, and there was a great diversity of opinions. They intend to evaluate the reasons for the students' resistance in future work.

In another study on the use of Programming Dojo in academic environments (DELGADO et al., 2012), the authors propose adaptations of Programming Dojo for use in higher education in the area of computing in order to create new forms of interaction between students and teachers. Adapted Dojos were applied in 3 different subjects, and the self-residents reported on the obstacles encountered and suggestions for getting around them. They put together a set of adaptations to the Programming Dojo to apply in undergraduate courses. Some features of Programming Dojo are relevant for teaching computer programming, but in undergraduate courses many subjects use programming as a tool to teach other concepts. Therefore, the process needs to be adapted and controlled in order to be used as a teaching activity. The authors considered the adaptations presented to be effective, but they are still at a stage of maturation. Overcoming the students' shyness was considered a challenge. Compiling problems for use in Programming Dojo sessions was suggested as future work.

1.3 CONTRIBUTION TO RELATED WORK

This research evaluated the opinions of participants in Programming Dojo, with different levels of experience, different motivations and different groups. The opinions of experts were evaluated through interviews which provided qualitative data on the activity. The results are presented in Section 4.1.1, Section 4.1.2 and Section 4.1.3. A discussion of the results is presented in Chapter 5. Contributing to knowledge about the use of the activity as a tool to help teach agile practices.

3 RESEARCH METHOD

During the research period, thirteen Programming Dojo meetings were held (DOJOPR, 2012), always taking care to record the minutes of the meetings with the opinion of the participants who highlighted positive and negative points according to their perception. This retrospective serves to evaluate and improve the meetings, and was used in the research to outline the questionnaires and interviews. An electronic questionnaire, Appendix 7.1, made available on the Internet and posted on the discussion lists of Brazilian Programming Dojo groups, was answered by Programming Dojo participants with different levels of experience. The intention of this questionnaire was to obtain the opinion of regular Programming Dojo participants about the agile practices used in the meetings. Semi-structured interviews were carried out, Appendix 7.5, with eight experienced Programming Dojo professionals who reported their opinions on the practice of the activity, providing the research with qualitative data on the Programming Dojo activity. Questionnaires were then administered, Appendices 7.2 and 7.3, to specialization students at the Federal Technological University of Paranà and volunteers from the Paranà Dojo community who had taken part in Programming Dojo meetings. The aim of these questionnaires was to obtain information from different groups, a group of students who participated compulsorily in the activity and a group of voluntary participants, in order to compare responses.

3.1 ELECTRONIC QUESTIONNAIRE

An electronic questionnaire, Appendix 7.1, was made available to volunteers participating in Programming Dojo from various regions of the country. Information on the profile of the participants was requested in order to compare the responses of the more experienced participants with those of the less experienced, considering those who had participated in more than 20 Programming Dojo meetings as experienced. The questionnaire asked for participants' opinions on the practice of Programming Dojo and its effects on learning agile development techniques. A Likert scale was used for the opinion questions.

3.2 INTERVIEWS

To better understand the subjective aspects of the Programming Dojo technique, semi-structured interviews, Appendix 7.4, were used to guide the interviewer and at the same time allow the interviewee to express their opinion. According to (KVALE, 1996), with interviews in qualitative research you try to understand something from the subject's point of view and discover the meaning of their experiences.

3.3 FACE-TO-FACE QUESTIONNAIRES

A questionnaire was used to determine the profile of the participants, Appendix 7.2, with questions related to professional experience and prior knowledge of agile method practices such as pair programming, baby steps and test-driven development. Another questionnaire obtained the participants' opinions on the Programming Dojo activity, Appendix 7.3. The questions were related to participation in the meeting held just before the questionnaire was completed. The questionnaires were sent to two different groups. One group was made up of specialization course students who had compulsorily participated in a Programming Dojo. The second group was made up of volunteers who attended a regular meeting of the Dojo Paranâ group. The questionnaire on the profile of the participants was presented before the activity and the questionnaire on the participants' opinions was presented after the activity.

The face-to-face questionnaires were carried out to help understand the difference in opinions about the Programming Dojo in different groups, with different motivation and experience.

4 RESULTS

We obtained 64 answers to the electronic questionnaire (Appendix 7.1), 8 answers from the interviews (Appendix 7.5), 21 answers from specialization students and 11 answers from participants in the Dojo Paranâ group meeting to the same questionnaires (Appendix 7.2 and Appendix 7.3).

4.1 DATA OBTAINED

In order to obtain the opinion of regular Programming Dojo participants, an electronic questionnaire was drawn up which was made available and sent to various communities that organize Programming Dojo meetings. The questionnaire identified the profile of the 64 volunteers who took part. Approximately 30% declared themselves to be professionals, just over 34% students and almost 36% professionals who were also studying.

The electronic questionnaire obtained information on three categories of questions. Information on the environment and techniques used, on the resources and infrastructure needed to carry out the activity and on the agile practices used in the meetings. With this information, it was possible to assess the opinions of members of the community of regular participants in Programming Dojo meetings about the environment, resources and agile practices. The detailed results are presented in Section 4.1.1.

Interviews with experts with experience in organizing Programming Dojo meetings were carried out and collected descriptive information. They made it possible to evaluate subjective aspects of the activity, presented in Section 4.1.2.

Identical face-to-face questionnaires were administered to two different groups. The information collected made it possible to evaluate and compare the opinions on the dynamics and agile practices of a group of regular participants in Programming Dojo and a group of specialization students from the Federal Technological University of Paraná. The group of students participated compulsorily during one of the classes and the regular participants did so voluntarily. The results are presented in Section 4.1.1.

4.1.1 Electronic questionnaire data

Questions were presented using a Likert scale, with five options: Strongly disagree (-2), Partially disagree (-1), Neutral (0), Partially agree (1) and Strongly agree (2). To improve the presentation of the results, the following legend was created:

- Q1 - Which category do you fall into?
- Q2 - How old are you?
- Q3 - How long have you been programming?
- Q4 - Gender:
- Q5 - Do you know what "Test Driven Development" is?
- Q6 - Experience with "Test Driven Development":

- Q7 - Experience with pair programming:
- Q8 - Do you know what pair programming is?
- Q9 - Do you know what baby steps are?
- Q10 - Experience with baby steps:
- Q11 - How long have you participated in program dojos?
- Q12 - How many programming dojo meetings have you attended?
- Q13 - How often do you participate in program dojos?
- Q14 - It's better when the time between meetings is shorter (for example, weekly is better than fortnightly).
- Q15 - Practice using programming languages that everyone knows results in better learning.
- Q16 - Works well when all participants are experienced.
- Q17 - Using the pair programming technique hinders the progress of the activity.
- Q18 - Baby steps hinder the process of finding a solution to the problem.
- Q19 - Creating the tests before the code helps the progress of the activity.
- Q20 - Computer labs are suitable for practicing programming dojo.
- Q21 - It is important that the program dojo takes place in a fixed location on a regular basis.
- Q22 - It is essential to have a board or flipchart for writing.
- Q23 - It's not necessary to have a projector if the pair communicates the idea well.
- Q24 - The programming dojo encourages learning new programming languages.
- Q25 - The programming dojo makes it difficult to learn TDD.
- Q26 - The programming dojo favors the leveling of participants due to pair programming.
- Q27 - Practicing baby steps teaches you to solve problems gradually.
- Q28 - During the programming dojo activity you learn a lot because of the exchange of experiences.
- Q29 - To learn agile methods, I prefer traditional training, with theoretical exposition and individual exercises.

Questions Q1 to Q13 asked for information on the profile of the participants. Questions Q14 to Q29 asked for participants' opinions using a Likert scale. Questions Q17, Q18, Q20, Q23, Q25, Q29 were reversed to avoid influencing participants to always answer positively.

Questions about the environment and techniques used

Questions Q14, Q15, Q16, Q17, Q18, Q19 displayed the following explanatory text to contextualize the question: "With regard to the *Randori-style* programming dojo activity (most commonly used style, doesn't require advance preparation, participants choose problem and/or language, has mechanism for swapping pairs, starts from scratch):", along with the question statement. The questions in this category were related to the environment, the preferred periodicity, the participants' level of experience, and the techniques and technologies used. Table 1 shows the distribution of the frequencies of responses in quantity and percentage for this group of questions.

Table 1: Frequency of responses to the electronic questionnaire about the Programming Dojo environment

	Strongly disagree		Partially disagree		Neutral		Partially agree		I totally agree	
	cnt	perc	cnt	perc	cnt	perc	cnt	perc	cnt	perc
Q14	4	6.2%	6	9.4%	15	23.4%	20	31.2%	19	29.7%
Q15	5	7.8%	9	14.1%	13	20.3%	26	40.6%	11	17.2%
Q16	12	18.8%	22	34.4%	8	12.5%	14	21.9%	8	12.5%
Q17	40	62.5%	14	21.9%	7	10.9%	3	4.7%	0	0.0%
Q18	34	53.1%	13	20.3%	9	14.1%	8	12.5%	0	0.0%
Q19	2	3.1%	4	6.2%	9	14.1%	17	26.6%	32	50.0%

Table 2 shows the mean, variance and standard deviation of the answers to the electronic questionnaire of the participants considered to be beginners. To be considered a beginner, the participant must have answered that they had been attending meetings for less than 11 months or claimed to have attended less than 10 Programming Dojo meetings.

Table 2: Mean, variance and standard deviation of the electronic questionnaire on the environment of the Programming Dojo - beginners

Variable	Average	Variance	Standard deviation
Q14	0.58	1.20	1.08
Q15	0.72	1.21	1.09
Q16	-0.12	1.87	1.35
Q17	-1.19	0.92	0.95
Q18	-0.91	1.32	1.14
Q19	1.09	1.04	1.01

Table 3 shows the mean, variance and standard deviation of the answers to the electronic questionnaire of the participants considered to be experienced. To be considered experienced, the participant must have answered that they have been attending meetings for more than 1 year and claimed to have attended 10 or more Programming Dojo meetings.

Figure 2 shows the frequency of responses to the electronic questionnaire, representing the trend of participants' opinions about the Programming Dojo environment.

Table 3: Mean, variance and standard deviation of the electronic questionnaire about the environment of the Programming Dojo - experienced

Variable	Average	Variance	Standard deviation
Q14	0.90	1.79	1.31
Q15	-0.10	1.29	1.11
Q16	-0.52	1.56	1.22
Q17	-1.90	0.09	0.29
Q18	-1.62	0.55	0.72
Q19	1.24	1.49	1.19

The graphs show the frequency of responses from all participants, experienced and inexperienced participants.

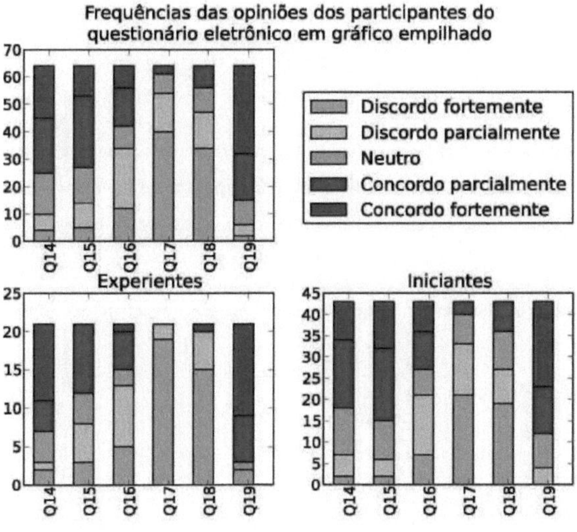

Figura 2: Frequency of responses to the electronic questionnaire on the environment of the Programming Dojo

Questions about the infrastructure and resources used

Questions Q20, Q21, Q22 and Q23 displayed the text: "With regard to the physical structure where programming dojo takes place:", along with the question statement. This category was related to the infrastructure and resources needed for the activity. Table 4 shows the distribution of the frequencies of responses in quantity and percentage for this group of questions.

Table 4: Frequency of responses to the electronic questionnaire on the infrastructure of the Programming Dojo

	Strongly disagree		Partially disagree		Neutral		Partially agree		I totally agree	
	cnt	perc	cnt	perc	cnt	perc	cnt	perc	cnt	perc

Q20	6	9.4%	13	20.3%	18	28.1%	11	17.2%	16	25.0%
Q21	4	6.2%	8	12.5%	17	26.6%	23	35.9%	12	18.8%
Q22	4	6.2%	3	4.7%	10	15.6%	22	34.4%	25	39.1%
Q23	30	46.9%	18	28.1%	4	6.2%	9	14.1%	3	4.7%

Table 5 shows the mean, variance and standard deviation of the answers to the electronic questionnaire of the participants considered to be beginners. To be considered a beginner, the participant must have answered that they had been attending meetings for less than 11 months or claimed to have attended less than 10 Programming Dojo meetings.

Table 5: Mean, variance and standard deviation of the electronic questionnaire on the infrastructure of the Programming Dojo - beginners

Variable	Average	Variance	Standard deviation
Q20	0.42	1.77	1.32
Q21	0.40	1.34	1.14
Q22	1.00	1.24	1.10
Q23	-0.95	1.52	1.22

Table 3 shows the mean, variance and standard deviation of the answers to the electronic questionnaire of the participants considered to be experienced. To be considered experienced, the participant must have answered that they have been attending meetings for more than 1 year and claimed to have attended 10 or more Programming Dojo meetings.

Table 6: Mean, variance and standard deviation of the electronic questionnaire on the infrastructure of the Programming Dojo - experienced

Variable	Average	Variance	Standard deviation
Q20	0.00	1.50	1.20
Q21	0.67	1.13	1.04
Q22	0.86	1.53	1.21
Q23	-1.05	1.65	1.25

Figure 3 shows the frequency of responses to the electronic questionnaire, representing the trend of participants' opinions on the infrastructure and resources used in the Programming Dojo. The graphs show the frequency of responses from all participants, experienced participants and inexperienced participants.

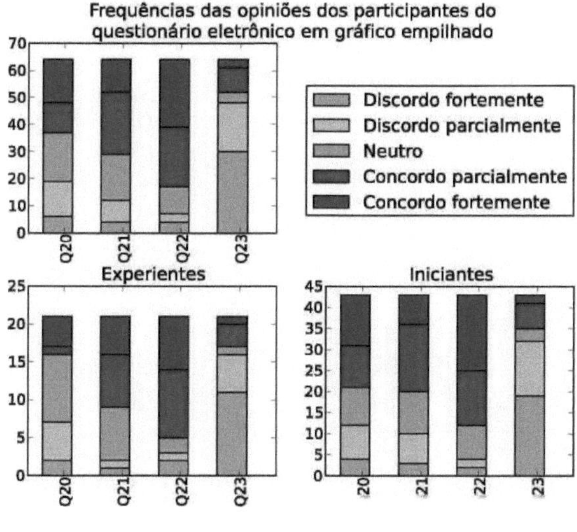

σ ○ o

Figura 3: Frequency of responses to the electronic questionnaire on the infrastructure used in the Programming Dojo

Questions about the perception of learning

Questions Q24, Q25, Q26, Q27, Q28, Q29 displayed the text: "Regarding your perception of learning during the programming dojo activity:", along with the question wording. These questions collected information on the participants' perception of learning various agile practices. Table 7 shows the frequency distribution of answers in quantity and percentage for this group of questions.

Table 8 shows the mean, variance and standard deviation of the answers to the electronic questionnaire of the participants considered to be beginners. To be considered a beginner, the participant must have answered that they had been attending meetings for less than 11 months or claimed to have attended less than 10 Programming Dojo meetings.

Table 9 shows the mean, variance and standard deviation of the answers to the electronic questionnaire of the participants considered experienced. To be considered experienced

Table 7: Frequency of responses to the electronic questionnaire on learning agile method practices used in the Programming Dojo

	Strongly disagree		Partially disagree		Neutral		Partially agree		I totally agree	
	cnt	perc	cnt	perc	cnt	perc	cnt	perc	cnt	perc
Q24	1	1.6%	2	3.1%	2	3.1%	20	31.2%	39	60.9%
Q25	37	57.8%	11	17.2%	10	15.6%	4	6.2%	2	3.1%
Q26	0	0.0%	5	7.8%	10	15.6%	26	40.6%	23	35.9%

Q27	1	1.6%	3	4.7%	6	9.4%	11	17.2%	43	67.2%
Q28	1	1.6%	2	3.1%	0	0.0%	13	20.3%	48	75.0%
Q29	26	40.6%	14	21.9%	16	25.0%	4	6.2%	4	6.2%

Table 8: Mean, variance and standard deviation of the electronic questionnaire on the perception of learning during the Programming Dojo - beginners

Variable	Average	Variance	Standard deviation
Q24	1.40	0.82	0.89
Q25	-1.05	1.33	1.14
Q26	0.91	0.99	0.98
Q27	1.26	1.19	1.08
Q28	1.51	0.83	0.90
Q29	-0.70	1.69	1.29

the participant must have answered that they have been attending meetings *for* more than 1 year and claimed to have attended 10 or more Programming Dojo meetings.

Table 9: Mean, variance and standard deviation of the electronic questionnaire on the perception of learning during the Programming Dojo - experienced

Variable	Average	Variance	Standard deviation
Q24	1.62	0.45	0.65
Q25	-1.52	0.96	0.96
Q26	1.33	0.43	0.64
Q27	1.81	0.16	0.39
Q28	1.90	0.09	0.29
Q29	-1.14	0.93	0.94

Figure 4 shows the frequency of responses to the electronic questionnaire, representing the trend of participants' opinions on their perception of learning agile practices in the Programming Dojo. The graphs show the frequency of responses from all participants, experienced participants and inexperienced participants.

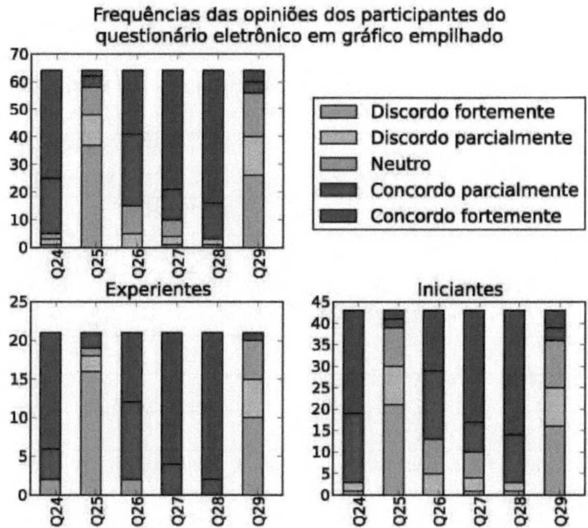

Figura 4: Frequency of responses to the electronic questionnaire on the perception of learning agile practices in the Programming Dojo

4.1.2 Interviews

Experts with extensive experience in Programming Dojo were invited to take part in an interview about Programming Dojo. The criterion adopted for the invitation was the fact that they had published a scientific article on Programming Dojo or had been nominated directly by the author of a scientific article on Programming Dojo. A summary of the interviews is available in Appendix 7.5. The interviews revealed the experts' opinions regarding the advantages and disadvantages of the activity, their personal experiences and situations where the Dojo can be used with more or less benefit. Table 10 shows the advantages and disadvantages according to the opinions of the interviewees.

The experts reported that the Programming Dojo increased participation and interaction between the students in their classrooms. The Programming Dojo also allowed classes to follow a rhythm that suited each student. According to one of the experts, the Programming Dojo makes it easier to clarify doubts thanks to the collaborative aspect of the activity.

On the other hand, several interviewees said that the Programming Dojo should intersperse its activities with presentations of theory, since the Programming Dojo focuses on

Table 10: Characteristics of the Programming Dojo according to experts

Advantages	Disadvantages
Collaborative and participative environment. Dynamic class. Increases communication.	Lectures present the broadest overview of concepts.
Suitable for practical activities. Focus on practicing a specific concept.	Focus on theory, on concepts.

Retrospective, a critical process of continuous improvement.	It can be prejudiced when carried out during working hours.
It can be used as a tool to assess candidates for job vacancies.	There may be little participation when it takes place outside of working hours.
The rhythm is determined by the participant using the keyboard, called the pilot.	Groups with frequent new members evolve very slowly.
Assiduous groups evolve gradually.	Environment exposes the participant.
It helps to lose inhibitions. Socializes participants.	Pressing the time limit for exchanging pairs.
Helps retain theory. Provides an opportunity to ask questions because everyone is watching.	It needs to be complemented by ex-positive lessons to introduce theory.
It strengthens the culture of agile practices.	
Encourages teamwork.	
More interactive than lectures.	

in practice. One of the experts said that sometimes the learner needs to be alone, in silence and try to do the activities on their own in order to grasp the content.

4.1.3 Data from face-to-face questionnaires

Face-to-face questionnaires were administered to two different groups. Each group answered two questionnaires, one about the profile of the participants and the other about the Programming Dojo activity. The first group of 22 people was made up of students on a specialization course at the Federal Technological University of Paranà, known as Group 1. The second group of 11 people was made up of volunteers from the Dojo Paranà community, known as Group 2. Group 1 participated in the Programming Dojo compulsorily during a class on the specialization course and Group 2 was made up of people interested in the Programming Dojo.

The questionnaires were given to the participants on different dates. While Group 1 had little knowledge of the activity, several members of Group 2 had already taken part in Programming Dojo meetings, as can be seen in Figure 5.

The aim of these questionnaires was to obtain the participants' opinions on the techniques used in the Programming Dojo, observing the differences and similarities in the responses of people with more and less experience in participating in the activity and people who participated compulsorily and voluntarily in the activity.

Figure 5 shows the experience in terms of number of Programming Dojo meetings claimed by the participants in the two groups. In Group 1, the vast majority, just over 90% of people, have never taken part in a Programming Dojo meeting. In Group 2, just over half, almost 55%, of the participants had never been to a meeting.

Experience with Programming Dojo

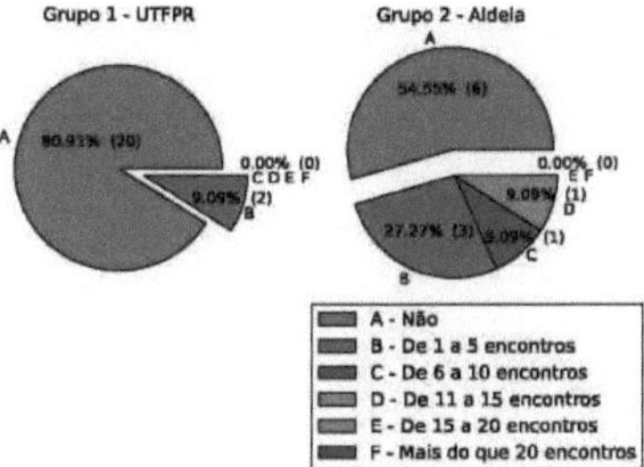

Figure 5: Participants' experience of Programming Dojo.

Figure 6 shows the results of the proportion of participants who said they knew what Programming Dojo was. In Group 1, less than 37% of the members said they knew about Programming Dojo. The majority of Group 2 participants, just under 82%, said they knew about Programming Dojo, although some said they had never taken part.

Previous knowledge of Programming Dojo

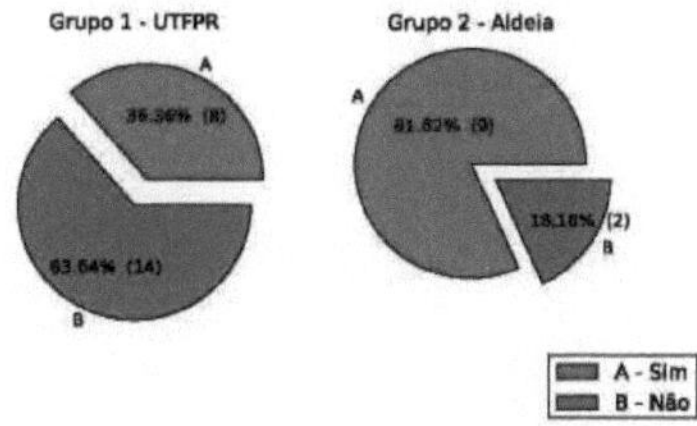

Figura 6: Participants' knowledge of Programming Dojo.

The questionnaire applied after the activity in both groups collected information on the participants' opinions. The questions were designed according to the results obtained from the electronic questionnaire in Section 4.1.1 and the interviews in Section 4.1.2, Appendix 7.5. The results of the questionnaire, which we refer to as the face-to-face questionnaire, made it possible to assess the opinions of two different groups, a group of volunteers and a group of students.

Questions were presented using a Likert scale, with five options: Strongly disagree (-2), Partially disagree (-1), Neutral (0), Partially agree (1) and Strongly agree (2). The same questions were presented to both groups. A legend was drawn up to improve the presentation of the results, as follows:

- Q1 - The programming dojo doesn't encourage student participation.
- Q2 - Communication of ideas is one of the strengths of the programming dojo.
- Q3 - Classes with the programming dojo are too slow.
- Q4 - Classes with the programming dojo are dynamic.

- Q5 - Programming dojos help programmers learn different languages.
- Q6 - There is more interaction between students when they attend dojos than when they attend lectures.
- Q7 - Classes with a programming dojo make it difficult to concentrate.

Q8 - The programming dojo helped me understand what TDD is.

Q9 - It's difficult to learn pair programming with the programming dojo.

- Q10 - I was able to understand the importance of baby steps during the programming dojo.

Questions Q1, Q3, Q7 and Q9 were reversed to avoid influencing participants to always answer positively. Table 11 shows the distribution of the frequencies of answers in quantity and percentage for Group 1, made up of students on a specialization course at the Federal Technological University of Paranà.

Table 11: Frequency of responses from Group 1 - UTFPR

	Strongly disagree	Partially disagree	Neutral	Partially agree	I totally agree
	cnt perc	cnt perc	cnt perc	cnt perc	cnt perc
Q1	10 47.6%	6 28.6%	1 4.8%	2 9.5%	2 9.5%
Q2	0 0.0%	3 14.3%	1 4.8%	9 42.9%	8 38.1%
Q3	2 9.5%	4 19.0%	4 19.0%	6 28.6%	5 23.8%
Q4	0 0.0%	3 14.3%	1 4.8%	8 38.1%	9 42.9%
Q5	0 0.0%	3 14.3%	0 0.0%	8 38.1%	10 47.6%
Q6	1 4.8%	1 4.8%	2 9.5%	9 42.9%	8 38.1%
Q7	1 4.8%	6 28.6%	2 9.5%	9 42.9%	3 14.3%
Q8	0 0.0%	1 4.8%	3 14.3%	11 52.4%	6 28.6%
Q9	4 19.0%	10 47.6%	1 4.8%	1 4.8%	5 23.8%
Q10	5 23.8%	1 4.8%	5 23.8%	4 19.0%	6 28.6%

The measures of dispersion, the mean, variance and standard deviation of the answers obtained from compulsory participants in the Programming Dojo held in a specialization course at the Federal Technological University of Paranà are shown in Table 12.

Table 12: Mean, variance and standard deviation of the questionnaire about the UTFPR Programming Dojo - Group 1

Variable	Average	Variance	Standard deviation
Q1	-0.95	1.85	1.33
Q2	1.05	1.05	1.00
Q3	0.38	1.75	1.29
Q4	1.10	1.09	1.02
Q5	1.19	1.06	1.01

Q6	1.05	1.15	1.05
Q7	0.33	1.43	1.17
Q8	1.05	0.65	0.79
Q9	-0.33	2.23	1.46
Q10	0.24	2.39	1.51

Figure 7 shows the frequency of responses from the participants in the Programming Dojo specialization class at the Federal Technological University of Paranà, representing the trend of the participants' opinions.

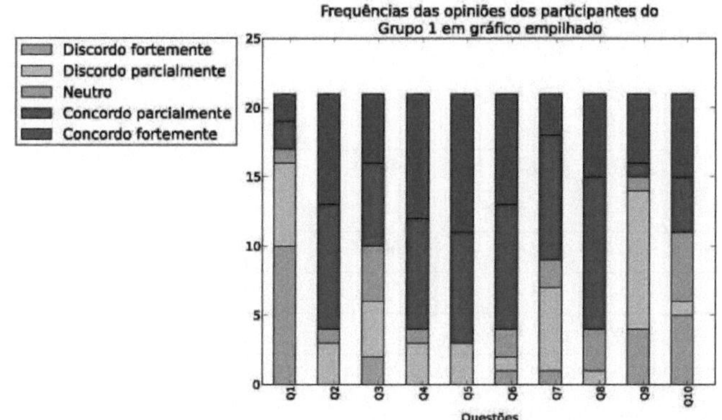

Figura 7: Frequency of responses from participants in the Programming Dojo at UTFPR, Group 1

Table 13 shows the distribution of the frequencies of responses in quantity and percentage for Group 2, made up of volunteer members of the Dojo Paranà community, who took part in a meeting in a room provided by Aldeia CoWorking, a collaborative workspace. One of the participants didn't answer question 5; this participant's answer was ignored in the statistical calculations for this question.

Table 13: Frequency of responses from Group 2 - CoWorking Village

	Strongly disagree	Partially disagree	Neutral	Partially agree	I totally agree
	cnt perc	cnt perc	cnt perc	cnt perc	cnt perc
Q1	9 81.8%	2 18.2%	0 0.0%	0 0.0%	0 0.0%
Q2	0 0.0%	0 0.0%	0 0.0%	1 9.1%	10 90.9%
Q3	3 27.3%	2 18.2%	4 36.4%	2 18.2%	0 0.0%
Q4	0 0.0%	0 0.0%	0 0.0%	2 18.2%	9 81.8%
Q5	0 0.0%	0 0.0%	0 0.0%	5 45.5%	5 45.5%
Q6	0 0.0%	0 0.0%	0 0.0%	4 36.4%	7 63.6%
Q7	4 36.4%	5 45.5%	1 9.1%	0 0.0%	1 9.1%
Q8	0 0.0%	0 0.0%	1 9.1%	4 36.4%	6 54.5%
Q9	6 54.5%	5 45.5%	0 0.0%	0 0.0%	0 0.0%
Q10	0 0.0%	1 9.1%	0 0.0%	4 36.4%	6 54.5%

The measures of dispersion, the mean, variance and standard deviation of the answers obtained from volunteer participants in the Programming Dojo held in a room provided by Aldeia CoWorking are shown in Table 12.

Table 14: Mean, variance and standard deviation of the questionnaire about the Village Programming Dojo - Group 2

Variable	Average	Variance	Standard deviation
Q1	-1.82	0.16	0.39
Q2	1.91	0.09	0.29
Q3	-0.55	1.27	1.08
Q4	1.82	0.16	0.39
Q5	1.50	0.28	0.50
Q6	1.64	0.25	0.48
Q7	-1.00	1.40	1.13
Q8	1.45	0.47	0.66
Q9	-1.55	0.27	0.50
Q10	1.36	0.85	0.88

Figure 8 shows the frequency of responses from the participants in the Programming Dojo at the CoWorking Village, representing the trend of the participants' opinions.

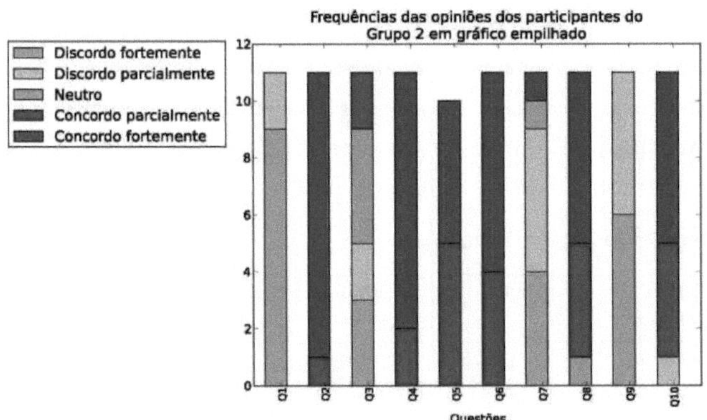

Figura 8: Frequency of responses from participants in the Village Programming Dojo, Group 2

4.1.4Comments on the data

The data collection took place between March 2011 and November 2012. Different groups took part in the survey: undergraduate students, specialization students, professionals who take part in regular Programming Dojo meetings and specialists with experience in organizing and conducting Programming Dojo meetings. The aim of the data survey was to identify the perception of several different profiles of people about the learning environment offered by Programming Dojo. Aspects such as the interaction between students, the dynamics and pace of the meeting as well as agile practices, o-driven development and testing using programming dojo, pair programming and baby steps, discussed in chapter 5, were evaluated.

The python programming language, the matplotlib library (HUNTER, 2007) and the Inkscape vector image editor were used to generate the graphs and statistical calculations.

5 DISCUSSION OF RESULTS

In a teaching environment, a classroom, whether it's a supplementary course in a training company or a university course, the whole environment matters. The resources used, the relationship between the participants, the way the concepts are presented all build up the learning experience.

In this way, the Programming Dojo was evaluated bearing in mind the need to look at the most diverse aspects of the learning experience. The electronic questionnaire 4.1.1 was evaluated by comparing the responses of experienced and novice participants. The interviews 4.1.2 were evaluated subjectively. The face-to-face questionnaires 4.1.3 were compared between compulsory participants, Group 1, and voluntary participants, Group 2.

5.0.5 Electronic questionnaire

The electronic questionnaire 4.1.1 was divided into categories to evaluate different aspects of the environment surrounding the Programming Dojo. The resources, infrastructure and agile method practices used were evaluated.

With regard to the environment and resources of the Programming Dojo, it was observed that the majority of participants agreed with Questions 14, 15 and 19, the majority disagreed with Question 16 and the majority also disagreed with Questions 17 and 18, which had their meanings reversed. The only question with a standard deviation below 1 was Question 17 (0.86), making it the question with the lowest dispersion of responses among the participants. When evaluating the responses of only the experienced participants, an even lower dispersion can be observed for Question 17 (0.29) and a relatively low dispersion for Question 18 (0.72), however the standard deviation for Question 14 increased (1.31 vs 1.17). This indicates that experienced participants have a greater dispersion of answers for Question 14. When analyzing the answers of beginner participants, we noticed small changes in the standard deviation compared to the standard deviation of all the answers. The influence of the answers from participants considered to be beginners is greater as there are more of them, 43 out of 64 are considered to be beginners and 21 are considered to be experienced.

In the category of questions related to the resources used, the answers presented the lowest averages, with a high dispersion, especially questions 20 and 21. Questions 22 and 23 showed a higher average, but still with a high dispersion. The analysis of the answers from beginners and experts shows little difference.

As for the questions about the agile practices used in the Programming Dojo, the answers had a high average, with a moderate standard deviation. The two questions with the highest standard deviation were 25 and 29. The answers from beginners and experts show some difference, except for question 25, for all the answers showed a standard deviation of (1.11), (0.96) for the answers from experts and (1.14) for the answers from beginners.

5.0.6 Interviews

The interviews with experts provided subjective information about the activity, allowing for a qualitative analysis of the Programming Dojo. The common terms extracted from the speeches and presented in Table 10 include the influence of the Programming Dojo on interaction between students, greater participation and teamwork. One consideration is repeated: interspersing lectures to present theoretical content with the Programming Dojo activity to fix the content by practicing.

5.0.7 Face-to-face questionnaire

Analyzing the dispersion of responses, Group 1 of compulsory participants shows high values for standard deviation 12, while Group 2 of voluntary participants shows low values for standard deviation 14 in the vast majority of questions. The exception in Group 1 is question 8, which had a standard deviation of less than 1 (0.79), and the exceptions in Group 2 are questions 3 and 7, which had a standard deviation of just over 1 (1.08 and 1.13).

In order to discuss the results of the face-to-face questionnaire, the percentages of partial and total agreement and the percentages of partial and total disagreement were grouped together, as shown in Table 15.

Table 15: Comparison of the grouped percentage of responses to the face-to-face questionnaire

	Group 1 - UTFPR			Group 2 - Village		
Question	Disagree	Neutral	Agree	Disagree	Neutral	Agree
Q1	76.2%	4.8%	19.0%	100.0%	0.0%	0.0%
Q2	14.3%	4.8%	81.0%	0.0%	0.0%	100.0%
Q3	28.5%	19.0%	52.4%	45.5%	36.4%	18.2%
Q4	14.3%	4.8%	81.0%	0.0%	0.0%	100.0%
Q5	14.3%	0.0%	85.7%	0.0%	0.0%	91.0%
Q6	9.6%	9.5%	81.0%	0.0%	0.0%	100.0%
Q7	33.4%	9.5%	57.2%	81.9%	9.1%	9.1%
Q8	4.8%	14.3%	81.0%	0.0%	9.1%	90.9%
Q9	66.6%	4.8%	28.6%	100.0%	0.0%	0.0%
Q10	28.6%	23.8%	47.6%	9.1%	0.0%	90.9%

Comparing the answers of the two groups, it is possible to see that most of the answers follow the same trend, except for a few questions. The majority of participants in Group 1 agree with question 3, that classes are slow with Programming Dojo, while the majority of participants in Group 2 disagree. The majority of participants in Group 1 agree with question 7, that Programming Dojo makes it difficult to concentrate, while the majority of participants in Group 2 disagree. Just under half of the participants in Group 1 agree with question 10, that the Dojo helps them realize the importance of baby steps, the vast majority in Group 2 also agree, but the dispersion of responses was high in Group 1, standard deviation 1.51, and low in Group 2, standard deviation 0.88.

5.0.8 Remarks

The response values show some differences from what was expected, according to the influence that the Programming Dojo could have on the participants. Most of the responses had a relatively high standard deviation, indicating a high degree of variation in the participants' opinions. The lowest standard deviation values were found in the responses of volunteer participants, Group 2 of the face-to-face questionnaire and former participants of the electronic questionnaire, indicating that participants come to have a different perception of the activity over time. Even with these differences, most of the questions followed the same trend of answers in all groups, the exceptions having been pointed out earlier in Chapter 4 and in this Chapter 5. The results of the answers also followed the opinions expressed by the experts in the interviews.

6 Conclusion

In short, the Programming Dojo can be used to improve the teaching of programming, especially agile practices, improving student participation and class dynamics. However, it should be interspersed with lectures to present concepts. It is also necessary to use tricks to help shy students lose their initial inhibitions. Most of the participants in the questionnaires claimed that the Programming Dojo helped them learn agile practices, pair programming, test-driven development and baby steps.

According to the experts' recurring reports, improvements in student participation and class dynamics have been observed. Also, according to the interviewees, the Dojo is a practical activity, ideal for fixing concepts, but it should be interspersed with theoretical lectures to present concepts. The shyness and motivation of the students were not assessed in this research, but in the articles (CARMO; BRAGANHOLO, 2012) and (DELGADO et al., 2012) the authors stated that students reported feeling embarrassed when taking part in the Programming Dojo. The answers with the lowest standard deviation values and consequently the lowest dispersion of opinions were the questions related to agile practices, as presented in Section 4.1.1 and Section 4.1.3. The majority of participants agreed with the statements in this category and disagreed with the statements that had their meaning reversed.

The information presented in this research can serve as a reference for those interested in adopting Programming Dojo to teach programming, agile practices and even practical activities in general. We also believe that there is a lack of studies evaluating cognitive aspects of learning using Programming Dojo.

Finally, we were able to see that academia is interested in the subject, given the opportunity to publish two articles in academic events related to education, one article (LUZ et al., 2012) at the Brazilian Symposium on Informatics in Education - SBIE 2013 and one article (LUZ et al., 2013) published at the *International Conference on Advanced Learning Technologies* - ICALT 2013.

6 REFERENCES

ACM International Collegiate Programming Contest. **ACM-ICPC Home page.** Accessed on: September 13, 2013. Available at: <http://icpc.baylor.edu/>.

ANICHE, M. F.; SILVEIRA, G. d. A. Increasing learning in an agile environment: lessons learned in an agile team. In: **2011 AGILE confe**

rence. Salt Lake City, UT, USA: [s.n.], 2011. p. 289-295. Available at: <http://ieeexplore.ieee.org/lpdocs/epic03/wrapper.htm?arnumber=6005834>.

BACHE, E. Test driven development: performing art. In: ABRAHAMSSON, P. et al. (Ed.). **Agile processes in software engineering and extreme programming**. Springer Berlin Heidelberg, 2009, (Lecture Notes in Business Information Processing, v. 31). p. 217-218. ISBN 978-3-642-01853-4. 10.1007/978-3-642-01853-4_38. Disponivel em: <http://dx.doi.org/10.1007/978-3-642-01853-4_38>.

BALIJEPALLY, V. et al. Are two heads better than one for software development? the productivity paradox of pair programming. **MIS Q.**, v. 33, p. 91-118, mar. 2009. ISSN 0276-7783. Available at: <http://dl.acm.org/citation.cfm?id=2017410.2017418>.

BECK. **Test driven development: by example**. Boston, MA, USA: Addison-Wesley Longman Publishing Co, Inc, 2002. ISBN 0321146530.

BECK, K.; ANDRES, C. **Extreme programming explained: embrace change (2nd edition)**. [S.l.]: Addison-Wesley Professional, 2004. ISBN 0321278658.

BOSSAVIT, L.; GAILLOT, E. **Coding dojo wiki: BabySteps**. Available at: <http://codingdojo.org/cgi-bin/wiki.pl?BabySteps>.

BOSSAVIT, L.; GAILLOT, E. The coder's dojo - a different way to teach and learn programming. In: BAUMEISTER, H.; MARCHESI, M.; HOLCOMBE, M. (Ed.). **Extreme programming and agile processes in software engineering**. Springer Berlin / Heidelberg, 2005, (Lecture Notes in Computer Science, v. 3556). p. 1156-1158. ISBN 978-3-540-26277-0. 10.1007/11499053_54. Available at: <http://dx.doi.org/10.1007/11499053_54>.

BOSSAVIT, L.; GAILLOT, E. **Paris Dojo**. dec. 2005. Http://codingdojo.org/cgi- bin/wiki.pl?ParisDojo. Available at: http://codin.gdojo.org/cgi-bin/wiki.pl? ParisDojo. Accessed on: September 26, 2013.

BRAVO, M.; GOLDMAN, A. Reinforcing the learning of agile practices using coding Dojos. In: SILLITTI, A. et al. (Ed.). **Agile processes in software engineering and extreme programming**. Springer Berlin Heidelberg, 2010, (Lecture Notes in Business Information Processing, v. 48). p. 379-380. ISBN 978-3-642-13054-0. 10.1007/978-3-642-13054-0_41. Disponivel em: <http://dx.doi.org/10.1007/978-3-642-13054-0_41>.

BRAVO, M. V. **Approaches for teaching extreme programming practices**. Dissertation (Master's Degree) - Institute of Mathematics and Statistics, University of Sao Paulo, May 2011. Available at:

<http://grenoble.ime.usp.br/ gold/orientados/dissertacao-MarianaBravo.pdf>.

BRUEGGE, B.; REISS, M.; SCHILLER, J. Agile principles in academic education: a case study. In: **2009 sixth international conference on information technology: new generations**. Las Vegas, NV, USA: [s.n.], 2009. p. 1684-1686. Available at: <http://ieeexplore.ieee.org/lpdocs/epic03/wrapper.htm?arnumber=5070901>.

CARMO, D.; BRAGANHOLO, V. A study on the didactic use of programming dojos. In: **Workshop on Computer Education. Brazilian Computer Society**. [S.l.: s.n.], 2012.

DELGADO, C.; TOLEDO, R. de; BRAGANHOLO, V. Use of dojos in computer science higher education. In: **Workshop on Computer Education (WEI). Brazilian Society of Computing**. [s.n.], 2012. ISSN 2175-2761. Available at: <http://www.imago.ufpr.br/csbc2012/anais_csbc/eventos/wei/artigos/Uso%2520de%2520Dojo s%2520no%

DOJOPR, G. Wiki, **Home dojo-parana/dojo-parana wiki**. Aug. 2012. Https://github.com/dojo-parana/dojo-parana/wiki/. Available at: <https://github.com/dojo- parana/dojo-parana/wiki/>.

ERICSSON, K. A.; KRAMPE, R. T.; TESCH-ROMER, C. The role of deliberate practice in the acquisition of expert performance. **Psychological review**, American Psychological Association, v. 100, n. 3, p. 363, 1993.

GAILLOT, E. **What is Coding Dojo**. September 2, 2012. Accessed on September 13, 2013. Available at: <http://codingdojo.org/cgi-bin/wiki.pl?WhatIsCodingDojo>.

HUNTER, J. D. Matplotlib: A 2d graphics environment. **Computing In Science & Engineering**, IEEE COMPUTER SOC, v. 9, n. 3, p. 90-95, 2007.

KOSTER, B. Agile methods fix software engineering course. **J. Comput. Small Coll.**, v. 22, p. 131â137, dec. 2006. ISSN 1937-4771. ACM. Available at: <http://dl.acm.org/citation.cfm?id=1181901.1181925>.

KVALE, S. **Interviews: an introduction to qualitative research inter viewing**. Sage Publications, 1996. ISBN 9780803958197. Available at: <http://books.google.com.br/books?id=tJPZAAAAMAAJ>.

LEDDY, E. Wiki, **PyClass - noisebridge**. 2012. Available at: <https://www.noisebridge.net/wiki/PyClass>.

LUZ, R.; NETO, A. G. S. S.; NORONHA, R. V. Using Programming Dojos to Teach Test-Driven Development. **Anais do Simpòsio Brasileiro de Informàtica na Educaçâo. ISSN 2316-6533**, Nov. 2012.

LUZ, R. B. d.; NETO, A. G. S. S.; NORONHA, R. V. Teaching tdd, the coding dojo style. In: **Advanced Learning Technologies (ICALT), 2013 IEEE 13th International Conference on**. [S.l.: s.n.], 2013. p. 371-375.

MELNIK, G.; MAURER, F. A cross-program investigation of students' perceptions of agile methods. In: **Proceedings of the 27th international conference on software engineering - ICSE '05**. St. Louis, MO,

USA: [s.n.], 2005. p. 481. Available at: <http://portal.acm.org/citation.cfm?doid=1062455.1062543>.

MILLER, J.; SMITH, M. A TDD approach to introducing students to embedded programming. In: **Proceedings of the 12th annual SIGCSE conference on innovation and technology in computer science education - ITiCSE '07**. Dundee, Scotland: [s.n.], 2007. p. 33. Available at: <http://portal.acm.org/citation.cfm?doid=1268784.1268797>.

PIMENTEL, E. P. et al. Continuous assessment of learning, skills and abilities in computer programming. In: **Anais do Workshop de Informàtica na Escola**. [S.l.: s.n.], 2003. v. 1,n. 1, p. 533-544.

PRINCE, M. Does active learning work? A review of the research. **JOURNAL OF ENGINEERING EDUCATION**, AMER SOC ENGINEERING EDUCATION, 1818 N ST, N W, STE 600, WASHINGTON, DC 20036 USA, v. 93, n. 3, jul. 2004. Available at: <http://www.konferenslund.se/pp/TAPPS_Prince.pdf>.

SATO, D. T.; CORBUCCI, H.; BRAVO, M. V. Coding dojo: an environment for learning and sharing agile practices. **AGILE Conference**, IEEE Computer Society, Los Alamitos, CA, USA, v. 0, p. 459-464, 2008.

SCHNEIDER, J. eXtreme programming - helpful or harmful in educating undergraduates? **Journal of Systems and Software**, v. 74, n. 2, p. 121-132, 2005. ISSN 01641212. Available at: <http://linkinghub.elsevier.com/retrieve/pii/S0164121203002929>.

SINIAALTO, M.; ABRAHAMSSON, P. A comparative case study on the impact of test-driven development on program design and test coverage. In: **Proceedings of the first international symposium on empirical software engineering and measurement**. Washington, DC, USA: IEEE Computer Society, 2007 (ESEM '07), p. 275-284. ISBN 0-7695-2886-4. Available at: <http://dx.doi.org/10.1109/ESEM.2007.2>.

SKIENA, S. S.; REVILLA, M. A. **Programming challenges: the programming contest training manual**. New York: Springer, 2003. ISBN 0387001638 9780387001630.

THOMAS, D. **CodeKata - How to Become a Better Developer**. 2001. Available at: http://codekata.pragprog.com/. Accessed on: April 18, 2011. Available at: <http://codekata.pragprog.com/>.

THOMAS, D. **CodeKata: codekata**. 28 Jan. 2007. Available at: <http://codekata.pragprog.com/2007/01/code_kata_backg.html>.

THOMAS, D. **CodeKata: code Kata-How it started**. 28 jul. 2007. Available at: <http://codekata.pragprog.com/2007/01/code_katahow_it.html>.

VODDE, B.; KOSKELA, L. Learning test-driven development by counting lines. **IEEE Software**, IEEE Computer Society, Los Alamitos, CA, USA, v. 24, p. 74-79, 2007. ISSN 07407459.

WILLIAMS, L.; KESSLER, R. **Pair Programming Illuminated**. Boston, MA, USA: Addison- Wesley Longman Publishing Co., Inc, 2002. ISBN 0201745763.

7 APPENDIX: QUESTIONNAIRES AND INTERVIEW SCRIPT WITH EXPERTS

The questions marked with [Likert Scale] had the following options:

- Strongly disagree
- Partially disagree
- Neutral
- Partially agree
- I totally agree

7.1 QUESTIONNAIRE OPEN TO THE COMMUNITY.

Program dojo learning quiz.

Many professionals and companies have adopted the programming dojo activity to teach and disseminate agile practices. Programming dojos have emerged as an activity for teaching programming to working professionals, targeting people who value their programming skills and have a strong motivation to improve themselves. This questionnaire is part of academic research at the Federal Technological University of Paranà into the practice of programming dojo.

1 Which category do you fall into?

( ) Student

( ) Professional

( ) Both

2 What is your professional (job/career) or academic (name of course) field of work? _______

3 How old are you? ____

4 How long have you been programming? ______________________________

5. gender:

( ) Male

( ) Female

6 Do you know what "Test Driven Development" is?

( ) Yes

()No

7 Experience with "Test Driven Development":

( ) Beginner

( ) Intermediate

( ) Experienced

8 Do you know what pair programming is?

( ) Yes

()No

9 Experience with pair programming:

( ) Beginner

( ) Intermediate

( ) Experienced

10 Do you know what baby steps are?

( ) Yes

( ) No

11 Experience with baby steps:

( ) Beginner

( ) Intermediate

( ) Experienced

12 How long have you been participating in program dojos?

13 How many programming dojo meetings have you been to?

14 How often do you participate in program dojos? *

( ) Weekly

( ) Fortnightly

( ) Monthly

( ) Once every two months

( ) Once every semester

15 Regarding the "Randori" style programming dojo activity (most commonly used style, no need to prepare in advance, participants choose problem and/or language, has mechanism for switching pairs, starts from scratch): [Likert scale]

- It's better when the time between meetings is shorter (for example, weekly is better than fortnightly).
- Practice using programming languages that everyone knows results in better learning.

- It works well when all the participants are experienced.
- Using the pair programming technique hinders the progress of the activity.
- Baby steps make it difficult to find a solution to the problem.
- Creating the tests before the code helps the progress of the activity.

16 Regarding the physical structure where the programming dojo takes place: [Likert scale].

- Computer labs are suitable for practicing programming dojo.
- It is important that the program dojo takes place in a fixed location on a regular basis.
- It is essential that there is a board or flip chart for writing.
- It's not necessary to have a projector if the duo communicates the idea well.

17 Regarding your perception of learning during the programming dojo activity: [Likert scale].

- The programming dojo encourages learning new programming languages.
- The programming dojo makes it difficult to learn TDD.
- The programming dojo favors the leveling of participants due to pair programming.
- Practicing baby steps teaches you to solve problems gradually.
- During the programming dojo activity you learn a lot because of the exchange of experiences.
- To learn agile methods, I prefer traditional training, with theoretical exposition and individual exercises.

18 Comments: Free field for additional comments.

72 FACE-TO-FACE QUESTIONNAIRE ON THE PROFILE OF THE PARTICIPANTS.

Dojo - UTFPR specialization - profile of participants

Questionnaire part of academic research for the Professional Master's Degree in Applied Computing at the Department of Informatics of the Federal Technological University of Paranà.

1 What is your main occupation? ____

2 What area do you work in? Name of position ____

3 What experience do you have in programming? Years worked.

4 Do you know what Test Driven Development / TDD is?

( ) Yes

()No

5. What experience do you have with TDD?

( ) None

( ) Less than 1 year

( ) More than 1 to 2 years

( ) More than 2 to 3 years

( ) More than 3 to 4 years

( ) More than 4 years

6. Do you know what pair programming is?

( ) Yes

( ) No

7 What experience do you have with pair programming?

( ) None

( ) Less than 1 year

( ) More than 1 to 2 years

( ) More than 2 to 3 years

( ) More than 3 to 4 years

( ) More than 4 years

8 Do you know what baby steps are? Baby steps in the context of software development.

( ) Yes

( ) No

9 What's your experience with baby steps?

( ) None

( ) Less than 1 year

( ) More than 1 to 2 years

( ) More than 2 to 3 years

( ) More than 3 to 4 years

( ) More than 4 years

10 Do you know any programming dojo?

( ) Yes

( ) No

11 Have you ever been to a programming dojo meeting?

( ) No

( ) From 1 to 5 meetings

( ) From 6 to 10 meetings

( ) From 11 to 15 meetings

( ) From 15 to 20 meetings

( ) More than 20 meetings

7.3 FACE-TO-FACE QUESTIONNAIRE ON PARTICIPANTS' OPINIONS.

Dojo - UTFPR Specialization - participants' opinions

Questionnaire part of academic research for the Professional Master's Degree in Applied Computing at the Department of Informatics of the Federal Technological University of Paranà.

1. **Answer the questions according to your opinion of your participation in the programming dojo.** [Likert scale]

- The programming dojo is not conducive to student participation.
- Communication of ideas is one of the strengths of the programming dojo.
- Programming dojo classes are very slow.
- Classes with the programming dojo are dynamic.
- Programming dojos help programmers learn different languages.
- There is more interaction between students when they attend dojos than when they attend lectures.
- Classes with a programming dojo make it difficult to concentrate.
- The programming dojo helped me understand what TDD is.
- It's difficult to learn pair programming with the programming dojo.
- I realized the importance of baby steps during the programming dojo.

7.4 INTERVIEW QUESTIONS

The interviews followed a semi-structured script, with the interviewees being free to express their opinions.

-Name.

-Profession.

- How did you hear about Dojo de Programa?

- What's your experience with Programming Dojo?
- In your opinion, what are the advantages of the Programming Dojo over lectures?
- In your opinion, what are the disadvantages of the Programming Dojo compared to lectures?
- Which agile practices are made easier to understand when taught with the Programming Dojo?
- Are there any of these practices that I would prefer to be taught in the Programming Dojo?
- Do you think any agile practice shouldn't be taught with Programming Dojo?
- Do you have any additional comments?

7.5 summary of interviews with program staff

Experts with extensive experience in Programming Dojo were interviewed. The experts chosen had published academic articles on Programming Dojo or were directly recommended by an expert who had published an academic article on Programming Dojo. The interviews are summarized below. The interviews are organized in alphabetical order.

7.5.1 Alexandre Freire

Alexandre Freire holds a master's degree in Computer Science from the Institute of Mathematics and Statistics - IME (USP). He works in software systems development and agile methods consultancy and training. He is a founding partner of Agilbits, a consultancy and development company.

He learned about the Programming Dojo in 2005 at the Extreme Programming - XP conference, held at the University of Sheffield, England, during the workshop presented by Laurent Bossavit and Emmanuel Gaillot The Coder's Dojo. When he returned to Brazil, together with Danilo Sato, Hugo Corbucci and Mariana Bravo, he started holding meetings at IME - USP.

During the interview, he highlighted some of the characteristics of the activity: repetition of exercises, motivation and learning new languages. As well as taking part in meetings at IME-USP, he also took part in meetings at the company where he worked.

He said that the activity brought personal growth through continuous practice. Regarding the baby steps, he commented that after a while this technique can be well understood by the apprentice in such a way as to allow them to stop using it in their day-to-day work in some cases.

Alexandre, when asked whether the Programming Dojo offers any advantages compared to lectures, said that one of the great advantages is participation, favoring communication and the exchange of ideas in a participatory activity.

As for his opinion on the disadvantage of the Programming Dojo compared to a lecture, he commented that he finds the Programming Dojo interesting because it focuses on specific aspects, whereas lectures can give a broad view without being restrictive like the Programming Dojo.

In lectures, the focus is on theory, the presentation of concepts, while in the Programming Dojo there is a strong focus on practice. According to Alexandre Freire, the Programming Dojo was designed to practice test-driven development (TDD). He also mentioned that the Dojo teaches retrospectives, which is part of the continuous improvement process. This activity *is* always carried out in the final minutes of the meeting where participants give their opinions on the positive and negative points of the meeting.

Asked what kind of teaching Dojo is not suitable for, Alexandre suggested that Programming Dojo is not suitable for learning a language that the participants don't know; it's more productive to have at least one language expert so that the activity brings useful results.

In his concluding remarks, Alexandre compared programming dojos to physical activity practices, suggesting that the encouragement of a group of friends helps to motivate both physical activity and programming practices.

7.5.2 Daniel Cukier

Daniel Cukier is technology director at Elo7, a marketplace for handicrafts, where artisans can sell their products. He got to know the Programming Dojo at IME-USP, with Danilo Sato and Hugo Corbucci. He then started coordinating a group at the company where he worked, Locaweb.

According to Daniel, the beginning was a bit difficult because people thought that Programming Dojo wasn't part of the job, people were afraid to participate for fear of being misunderstood by their superiors and teammates.

When they tried an alternative, doing it outside of working hours, they ran into another difficulty: people had already left and didn't show up to take part. Little by little, the activity gained followers and support from the management, who invited and encouraged employees to take part via e-mail.

Then the meetings were opened up to the community, meaning that people who didn't work at Locaweb could take part. At this point, the Programming Dojo began to be more successful, but it became dependent on one person organizing the meeting, sending out invitations and taking the lead.

According to Daniel, the meetings stopped happening when he stopped organizing them. He said that in order for the Programming Dojo to succeed, there needs to be that person who embraces the idea and makes it happen.

Asked about the results of the meetings, he compared the IME-USP group with the Locaweb group. At IME-USP, the group meets regularly and most of the members are permanent, so they have been able to deepen concepts gradually.

In the Locaweb group, due to the particular characteristics, the size of the company and the progress of the meetings, new members were arriving, so it was necessary to start the Programming Dojo at a basic, introductory level.

He also reported that, even with this particularity, after around 20 meetings, the participants were able to understand what TDD is in practice. He said that the Programming Dojo had another outstanding aspect: it socialized development, strengthened the culture of pair programming and helped to humanize the programming activity.

He commented on variations of Programming Dojo: Dojo for image generation where testing is not automated but visual, Dojo for virtual machines with service testing and the Kake format where several pairs program in parallel with different problems and/or languages.

He also suggested that the Programming Dojo could be a good training tool. At Locaweb, some participants in the Programming Dojo meetings have been con- trolled.

According to Daniel, there are two important things in the Programming Dojo: the retrospective, which allows reflection on the activity, and the food, which allows the participants to socialize and exchange information more spontaneously.

He finished by commenting that the Dojo encourages teamwork and interaction between people, two important characteristics in companies.

7.5.3 Daniel Wildt

Daniel Wildt has been working in software development *for* over 15 years. He worked as a university lecturer for over 7 years. He started using Programming Dojo in the academic environment in 2005. During his Java programming classes, he noticed that the students couldn't keep up with the content.

He initially used pair programming, but it happened that some pairs were more engaged than others. He then began to work with a machine and just one pair, with the rest of the students remaining around the machine and changing pairs every 5 minutes.

In their classes, the first part was expository with some theory and the second part used the Programming Dojo activity. They used different forms of Programming Dojo: *Prepared*, *Randori* and *Kake*.

According to Daniel Wildt, one advantage of this activity is that, from the very first lesson, automated tests were used instead of using the console output to test the program. Daniel Wildt then noticed greater engagement from the students, who began to participate more in class and were able to show their qualities on a daily basis. Some received referrals for professional opportunities.

When asked about the continuity of the Programming Dojo, he said that after he left the faculty, one student continued, but participation decreased and at the time of the interview the Programming Dojo at the faculty was no longer taking place.

According to Daniel Wildt, the Programming Dojo facilitates the understanding of some practices, techniques and technologies, among them simple *design*, pace of work, test automation, object orientation and design patterns, introduction to new technologies such as the Ruby language, Javascript language.

Another point mentioned by Daniel Wildt was the mix of technologies. In a Java course, he held a Programming Dojo in the *Kake* format, with three computers and three pairs, the same problem and three different languages, inciting curiosity and showing the diversity of options available to professionals.

7.5.4 Danilo Sato

Danilo Sato is currently a consultant at Thoughtworks. He got to know Programming Dojo in 2006, during a workshop organized by Emilie Bachè during the Agile Processes in Software Engineering and Extreme Programming event. When he returned to Brazil, he started meeting at IME-USP.

According to Danilo, the Programming Dojo is more interactive than lectures. During the Programming Dojo, everyone solves the exercise at the same time. According to Danilo, the experience of building the solution on top of what others have done helps to retain the content.

He mentioned TDD and pair programming as agile practices for which the Programming Dojo favors teaching. Danilo commented that occasionally, when he can't attend Programming Dojo meetings, he uses some Programming Dojo techniques on his own. Obviously he doesn't do pair programming, but he uses TDD and baby steps to solve challenges while passing the time during a plane trip.

7.5.5 Elizabeth Leddy

Elizabeth Leddy, a software consultant and member of the development team for Plone, a free content management system, gave an unstructured interview. The author found it interesting to interview her because of her experience running programming Dojos in her classes.

Elizabeth is a volunteer at a *hackerspace* called Noisebridge in San Francisco, California, where she teaches Python programming (LEDDY, 2012) to people who want to change careers and develop software for the internet. Classes have been running since 2011. At the time of the interview, in 2012, there had already been three classes, approximately 50% of the participants had been there since the beginning and the other 50% varied with each new class.

Elizabeth said that, at first, during the lessons, some students understood the subject quickly and some didn't, but instead of asking questions, they stayed quiet.

After learning about the Programming Dojo, she tried it out during her classes and commented that it quickly became extremely popular. The topic taught is particularly complex: teaching web development to people who don't know how to program and who want to change careers. These people need to learn practical concepts so that they can apply them to personal projects at a reasonable pace, without rushing.

The Programming Dojo has helped them with this because the class follows the rhythm of the person who is using the computer, called a pilot. Another relevant point is that people interact more, ask each other more questions and create bonds that extend into everyday life after class. Some students have started personal projects together.

The structure of the classes is weekly and organized in such a way that each week is interspersed with a class to present concepts and a practical class with the Programming Dojo.

According to Elizabeth, the Programming Dojo makes the class more dynamic. Occasionally, when the pair using the computer can't solve something, a student with a little more knowledge is invited and what happens is that this student, because of their confidence and security in the subject, talks a lot and helps the group along, making the environment participatory.

7.5.6 Hugo Corbucci

Hugo Corbucci is a programmer and consultant, currently working at Thoughtworks. He co-created the Programming Dojo with Danilo Sato in 2006, who was returning from the Agile Processes in Software Engineering and Extreme Programming conference and organized the first meeting at the Institute of Mathematics and Statistics of the University of Sâo Paulo - IME-USP.

They continued to participate in meetings at IME-USP, and took part in other meetings: in Grenoble and Paris in France, and some in the United States. They had a paper accepted for the Agile Conference in 2008 (SATO et al., 2008).

According to Hugo, the Programming Dojo offers a collaborative environment that favors participation, allowing each participant to make the most of the learning experience.

Asked about the disadvantages of the Programming Dojo, he said that there are times when it is necessary to be silent, alone and reflect without the pressure of time, used in the Programming Dojo for peer exchange and without the pressure of the audience watching. This allows the student to absorb some types of content better.

Programming Dojo is more suitable for practice. As agile practices that Programming Dojo facilitates learning, he cited: pair programming, unit testing, TDD after learning unit testing and incremental development. The latter, facilitated by Programming Dojo, helps participants realize that it is possible to take one step at a time without having to solve the whole problem at once.

He finished by commenting that he had noticed a growth in interest in the Programming Dojo. The Dojo Sâo Paulo mailing list has many participants and there is a growing interest among people in practicing and continuing to learn.

7.5.7 Mauricio Aniche

Mauricio Aniche is a consultant and instructor at Caelum. He learned about Dojo de Programaçâo at IME-USP. He has attended meetings at Locaweb, Caelum and elsewhere.

According to him, Programming Dojo is a tool that helps participants lose their inhibitions, a practice that helps them socialize.

Asked about the comparison between lectures and Programming Dojo, he said that Programming Dojo is very practical. Although it's not a substitute for lectures, it can be used together.

Mauricio said that the Programming Dojo needs to be guided in order to be useful. The help of someone who is an expert in the subject or in the programming language chosen for the Programming Dojo favors learning.

For Mauricio, the Programming Dojo is especially suitable for teaching pair programming. If it's well guided, it's interesting for teaching TDD and design practices and using continuous integration tools as well as the discipline needed for continuous integration.

He also recounted his experience at Caelum, where Programming Dojo was used by the team of instructors.

After a while, the team lost interest in Programming Dojo because they couldn't get to the end of solving relatively simple problems. They decided to replace Programming Dojo with technical lectures. They tried using Programming Dojo in a training session, but haven't yet been able to evaluate the results.

He believes that the Programming Dojo can be used to teach programming both in universities and in training centers. He commented that he believes the Programming Dojo can be improved if it has clear objectives and guidelines.

7.5.8 Roberto Rodrigues

Roberto Rodrigues is a student at the School of Arts, Sciences and Humanities at the University of Sâo Paulo (EACH-USP) and works at the Free Software Competence Center at the University of Sâo Paulo (CCSL-USP). He researches usability and agile methodologies at the university.

Learned about the Programming Dojo at IME-USP. Organizer of the Programming Dojos at EACH-USP. In the interview, he highlighted the collaborative aspect of the Programming Dojo. According to him, even experienced people have a lot to learn during Programming Dojo meetings.

When asked about the relationship between Programming Dojo and lectures, he said that Programming Dojo favors learning in practice, because it's common for students to believe they've understood the theory and then realize when they try to do it themselves that they haven't. In Programming Dojo they have the chance to ask questions immediately, as everyone is watching and willing to help. In the Programming Dojo, they have the chance to clear up their doubts immediately, as everyone is watching and willing to help, whereas in lectures they don't always take their doubts to the teacher or instructor straight away.

He reported that he had not yet achieved good results with students who knew nothing and learned to program exclusively with Programming Dojo, but said that the learning curve for students who already knew the basics of programming is very fast.

A disadvantage of the Programming Dojo, according to Roberto, is that it focuses too much on practice, thus losing some of the concept, and it is necessary to use additional artifices to solve this, combining some activity to cover the theoretical part that the Programming Dojo lacks.

According to him, the Programming Dojo is especially suitable for teaching practices that require discipline or are very repetitive, such as TDD, continuous integration and retrospective.

He finished by commenting that the Programming Dojo needs a theoretical introduction to the techniques used to improve the results of the activity, and he thinks it's essential for the instructor to make clear the importance of answering questions during the activity.

7.5.9 Considerations on the interviews

During the interviews, it was possible to see that the experts share some similar opinions. The focus on practical activities, interspersing the activity with theoretical presentations, greater interaction between the students, are passive points among the experts.

However, some characteristics divide opinion. One of the experts says that experimenting alone without the pressure of time and an observing audience helps to fix certain concepts. Another expert says that the Programming Dojo helps students lose their inhibitions. While one expert says that the Dojo is not suitable for learning a new language, another says that the environment of the Programming Dojo is ideal for learning new languages.

Comparing these opinions with articles (DELGADO et al., 2012) and (CARMO; BRAGA- NHOLO, 2012), we found that similar behaviors were observed. The articles reported shyness, insecurity and a lack of interest in taking part, requiring incentives. The articles did not mention a focus on practical activities.

Printed by Books on Demand GmbH, Norderstedt / Germany